AF710563

THE ANARCHIST HANDBOOK

Robert Wells

J. FLORES
PUBLICATIONS
P.O. BOX 830131
MIAMI, FL 33283-0131

THE ANARCHIST HANDBOOK

Published by:
J. Flores Publications
P.O. Box 163001
Miami, FL 33116

Direct inquires and/or order to the above address.

Neither the author nor the publisher assumes any responsibility for the use or misuse of the information contained in this book. The author and publisher specifically disclaim any personal liability, loss, or risk incurred as a consequence of the use and application, either directly or indirectly, of any advice or information presented herein.

Technical data presented herein, particularly technical data on dealing with dangerous chemicals, poisons, explosives, drugs, and safety procedures inevitably reflects the author's individual beliefs and experience with particular equipment and environments under specific circumstances which the reader cannot duplicate or experience exactly. The information in this book should therefore be used for guidance only and should be approached with great caution.

Since neither the author nor the publisher have control over the materials used, use of techniques or equipment, or the abilities of the reader, no responsibility, either implied or expressed, is assumed for the use or misuse of the data or procedures given in this book.

ISBN 0-918751-02-0

Printed in the United States of America

WARNING!

Neither the author nor the publisher assumes any responsibility for the use or misuse of the information contained in this book. The author and publisher specifically disclaim any personal liability, loss, or risk incurred as a consequence of the use and application, either directly or indirectly, of any advice or information presented herein.

Technical data presented herein, particularly technical data on dealing with dangerous chemicals, poisons, explosives, drugs, and safety procedures inevitably reflects the author's individual beliefs and experience with particular equipment and environments under specific circumstances which the reader cannot duplicate or experience exactly. The information in this book should therefore be used for guidance only and should be approached with great caution.

Since neither the author nor the publisher have control over the materials used, use of techniques or equipment, or the abilities of the reader, no responsibility, either implied or expressed, is assumed for the use or misuse of the data or procedures given in this book.

CONTENTS

Introduction

This manual has been written to provide information on the safe construction of improvised weapons and munitions to military and law enforcement personnel, as well as survivalists and weapon enthusiasts.

The devices presented can be easily constructed from materials available to anyone from drug or hardware stores, hobby shops, supermarkets, or scrounged from junk piles.

The book containes little which could be considered new information, because it has drawn heavily on devices and techniques which have been well proved out, and are practical, versatile, and not too difficult to do.

Keep in mind that unless a special license is obtained from the Federal, State, and local governments, all devices presented are illegal to construct and to own. This information is presented as reference only or for possible future use, when having the knowledge contained in these pages may well spell the difference between life and death.

Expedient Silencer

Silencers for small arms can be made from steel gas or water pipe and fittings.

MATERIAL REQUIRED:

Aluminum or plastic container (soda pop can, shampoo bottle, etc.)

Steel pipe nipple, 6 inch long—See Table 1 for diameter.

Two steel pipe couplings—See Table 2 for dimensions

Cotton cloth—See Table 2 for dimensions

Drill

Absorbent cotton or copper scouring pads (more effective)

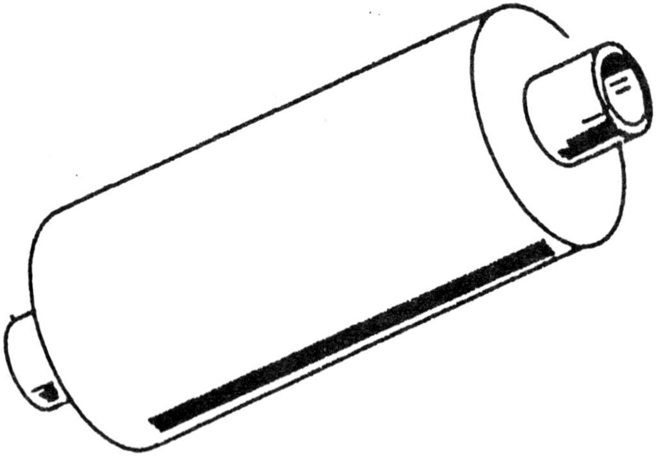

PROCEDURE:

1) Drill hole in container at both ends to fit outside diameter of pipe nipple. (See Table 1.)

2) Drill four (4) rows of holes in pipe nipple. Use Table 1 for diameter and location of holes.

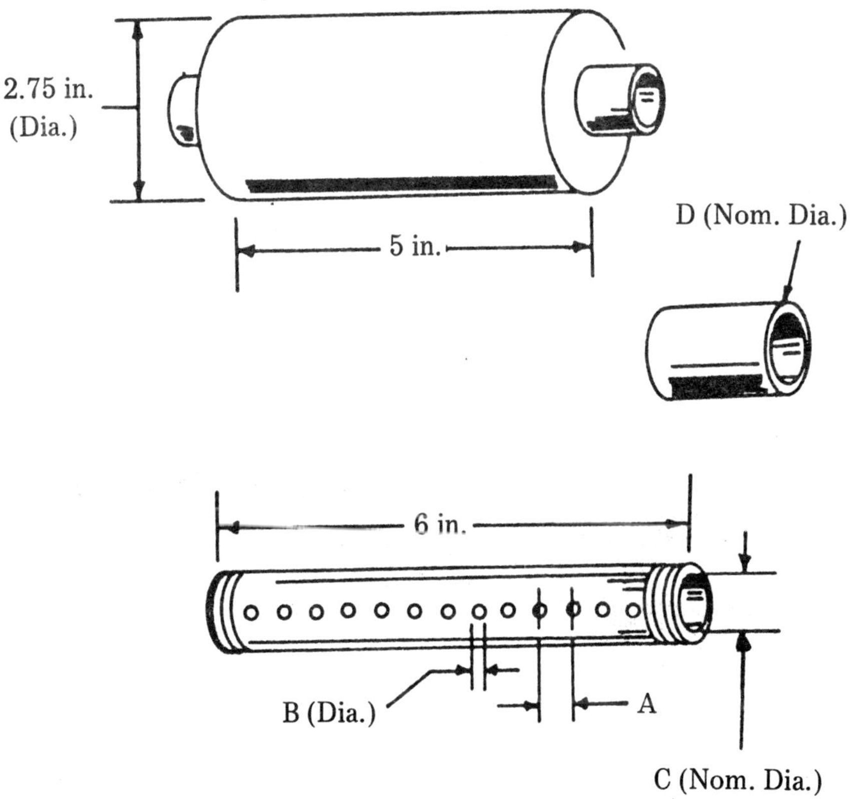

Table 1. Suppressor System Dimensions

	A	B	C	Coupling D	Holes per Row	(4-Rows) Total
.45 Cal.	3/8	1/4	3/8	3/8	12	48
.38 Cal.	3/8	1/4	1/4	1/4	12	48
9mm	3/8	1/4	1/4	1/4	12	48
7.62mm	3/8	1/4	1/4	1/4	12	48
.22 Cal.	1/4	5/32	1/8*	1/8	14	50

*Extra Heavy Pipe
All dimensions in inches

3) Thread one of the pipe couplings on the drilled nipple.

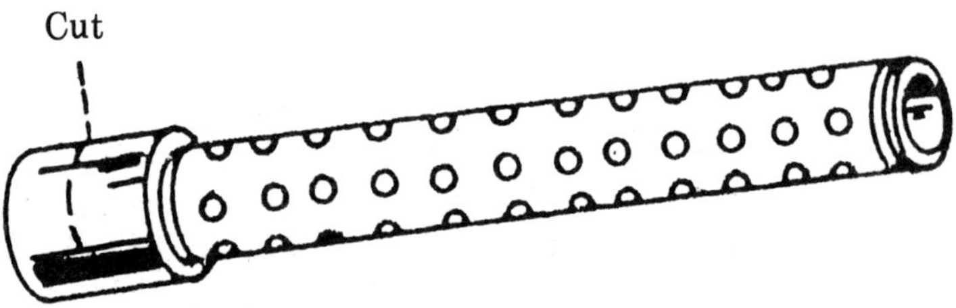

4) Cut coupling length to allow barrel of weapon to thread fully into suppressor system. Barrel should butt against end of the drilled pipe nipple.

5) Separate the top half of the container from the bottom half.

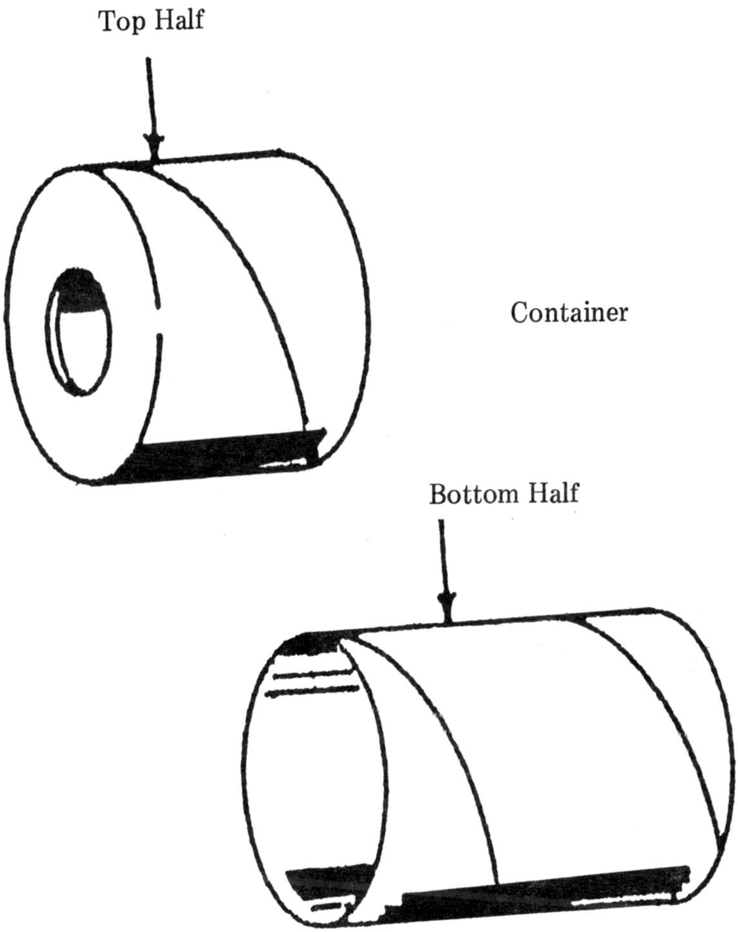

6) Insert the pipe nipple in the drilled hole at the base of the bottom half of the container. Pack the absorbent cotton inside the container and around the pipe nipple.

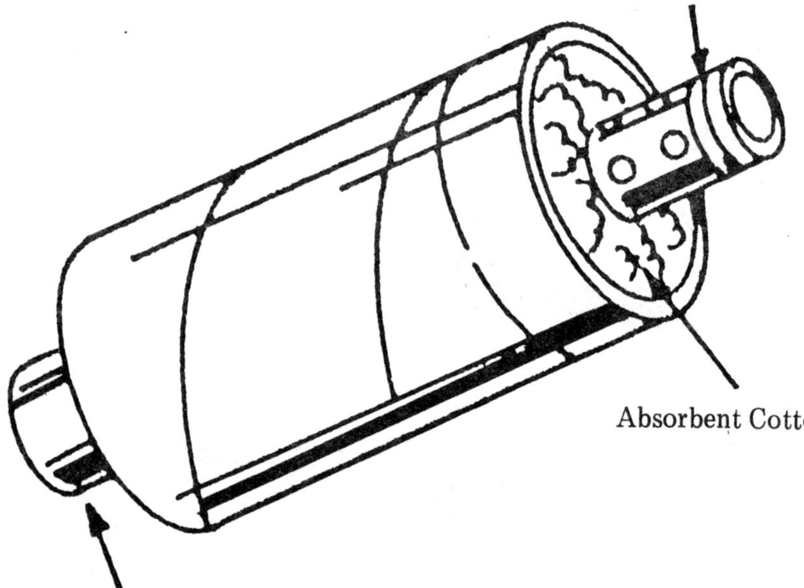

7) Pack the absorbent cotton in the top half of the container leaving a hole in the center. Assemble container to the bottom half.

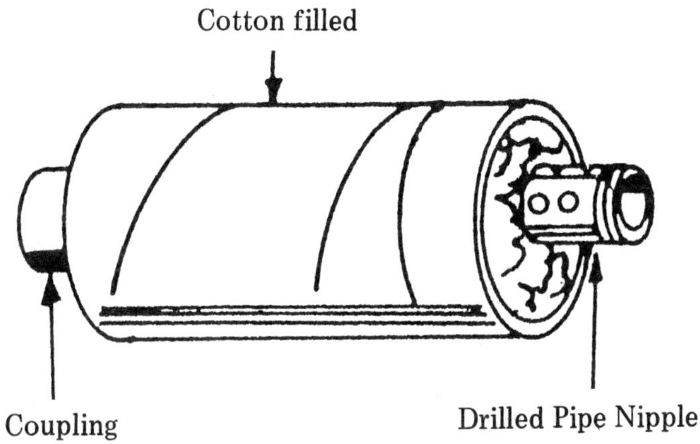

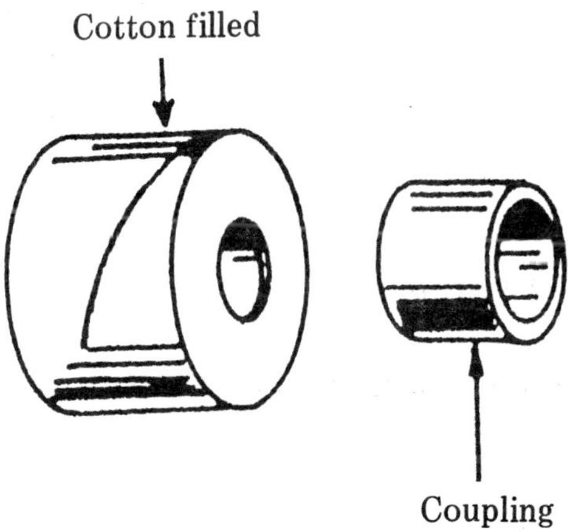

8) Thread the other coupling onto the pipe nipple.

NOTE: A longer container and pipe nipple, with same "A" and "B" dimensions as those given, will further reduce the sound of the system.

HOW TO USE:

1) Thread the suppressing system on the selected weapon securely.

2) Place the proper cotton wad size into the muzzle end of the system.

Table 2. Cotton Wadding – Sizes

Weapon	Cotton Wad Size
.45 Cal.	1½ x 6 inches
.38 Cal.	1 x 4 inches
9mm	1 x 4 inches
7.62mm	1 x 4 inches
.22 Cal.	Not needed

3) Load weapon

4) Weapon is now ready for use.

Improvised Blasting Caps

Mortars, mines and similar weapons often make use of electric initiators. An electric initiator can be made using a flashlight or automobile electric light bulb.

MATERIALS REQUIRED:

Electric light bulb and mating socket

Cardboard or heavy paper

Black powder

Adhesive tape

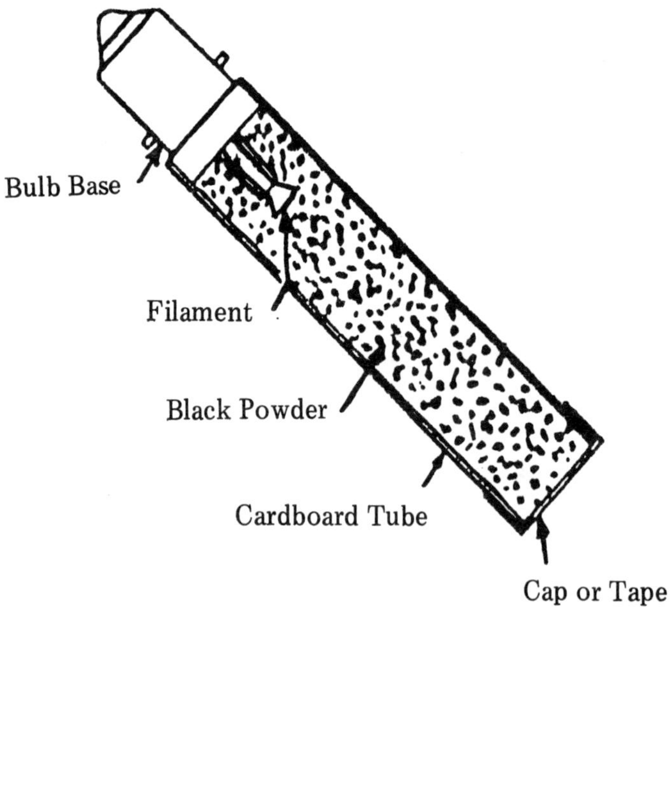

PROCEDURE:

METHOD 1

1) Break the glass of the electric light bulb. Take care not to damage the filament. The initiator will NOT work if the filament is broken. Remove all glass above the base of the bulb.

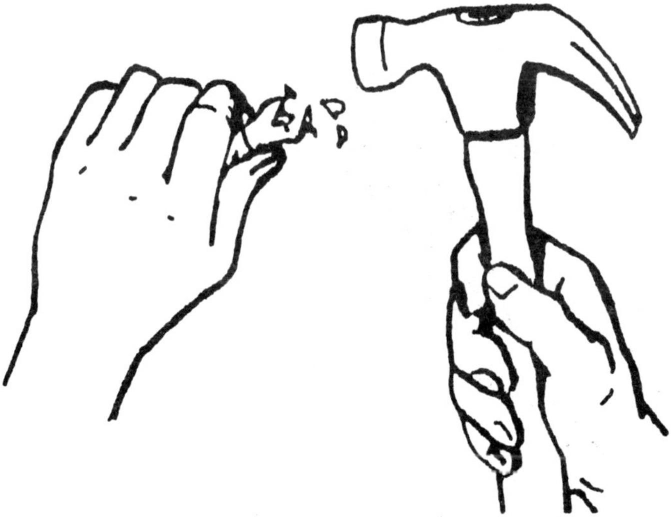

2) Form a tube 3 to 4 inches long from cardboard or heavy paper to fit around the base of the bulb. Join the tube with adhesive tape.

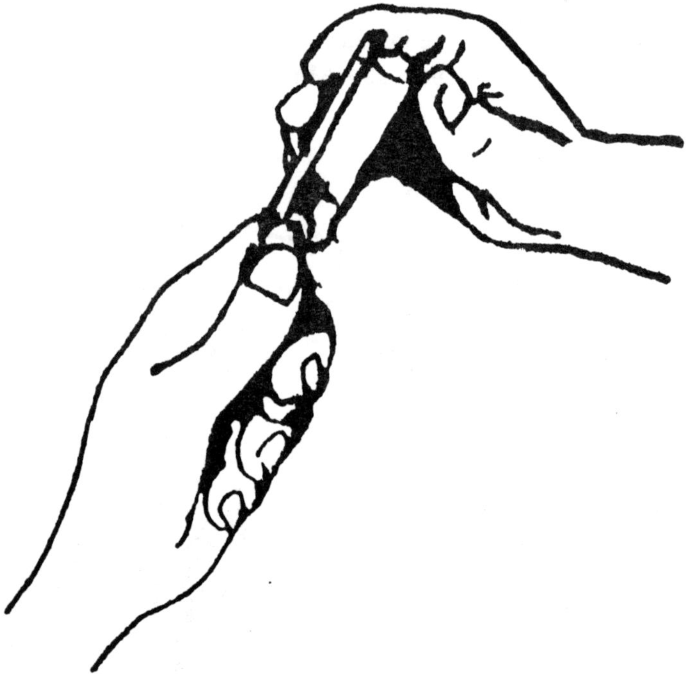

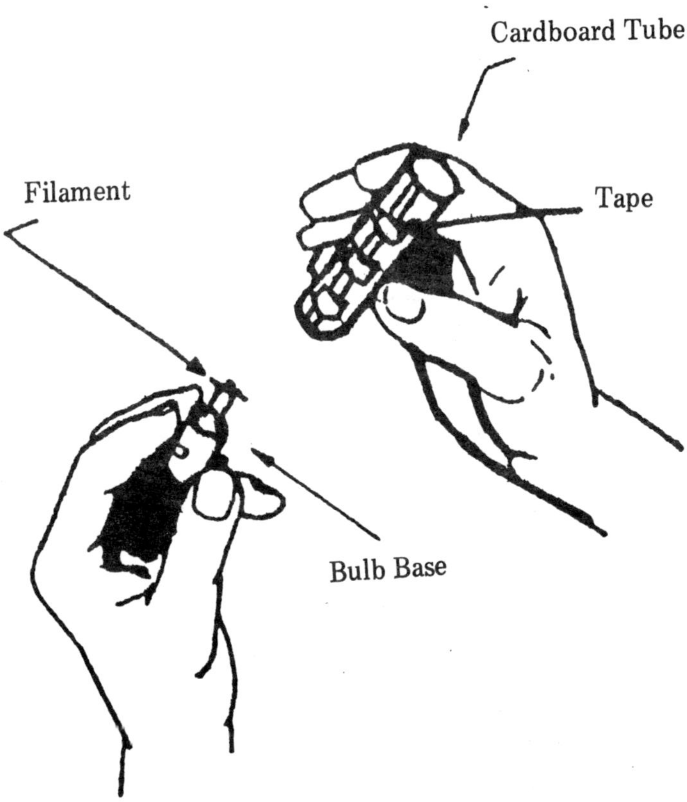

3) Fit the tube to the bulb base and tape in place. Make sure that the tube does not cover that portion of the bulb base that fits into the socket.

4) If no socket is available for connecting the initiator to the firing circuit, solder the connecting wires to the bulb base.

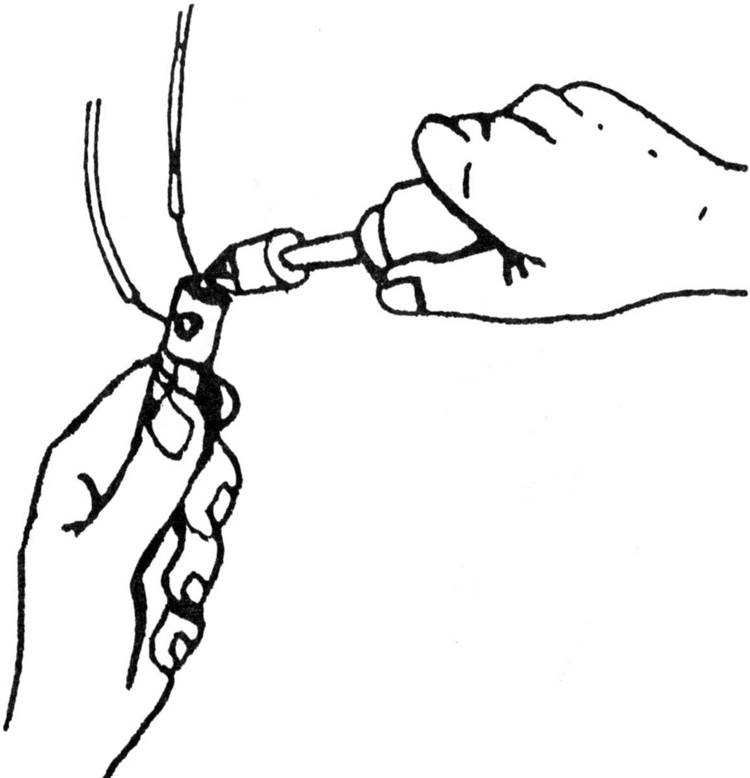

CAUTION: Do NOT use a hot soldering iron on the completed igniter since it may ignite the Black Powder.

5) Fill the tube with Black Powder and tape the open end of the tube closed.

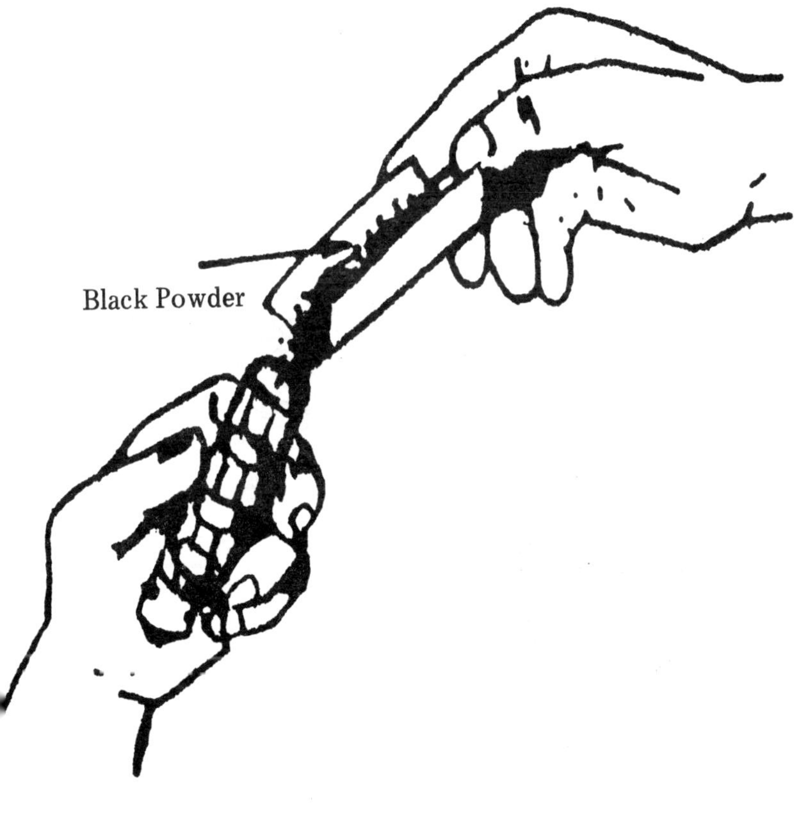

METHOD 2

If the glass bulb (electric light) is large enough to hold the Black Powder, it can be used as the container.

PROCEDURE:

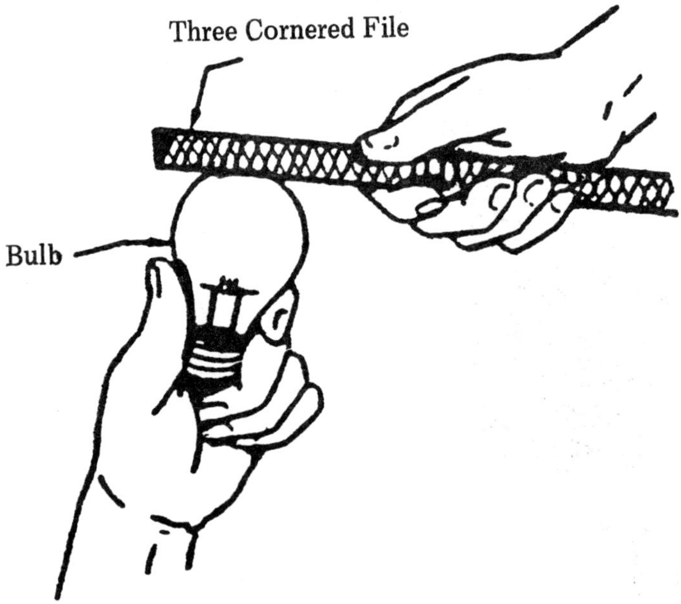

1) File a small hole in the top of the bulb.

2) Fill the bulb with Black Powder and tape the hole closed.

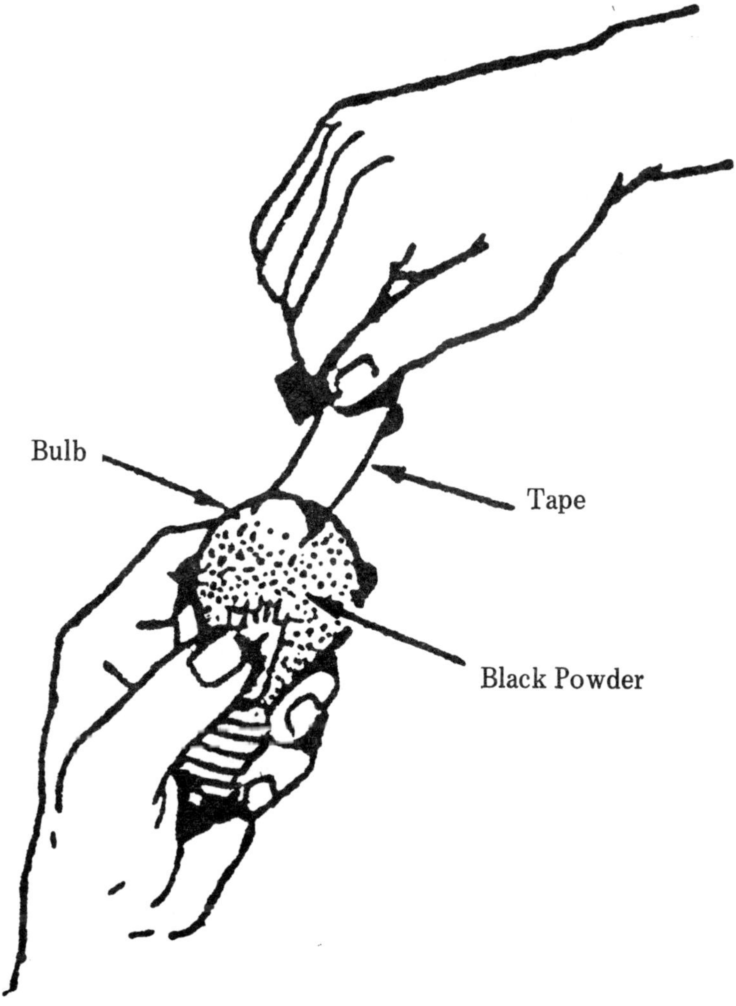

NON-ELECTRIC DETONATOR (BLASTING CAP)

Detonators (blasting caps) can be made from a used small arms cartridge case and field manufactured explosives.

MATERIAL REQUIRED:

- Black Powder
- Used cartridge case (.22 caliber or larger)
- Fuse, 12 inch long
- Round wooden stick (small enough just to fit in the neck of the cartridge case)
- Drill or knife
- Long nail with sharpened end
- Vise
- Improvised loading fixture

PROCEDURE:

1) Remove fired primer from a used cartridge case using a sharpened nail.

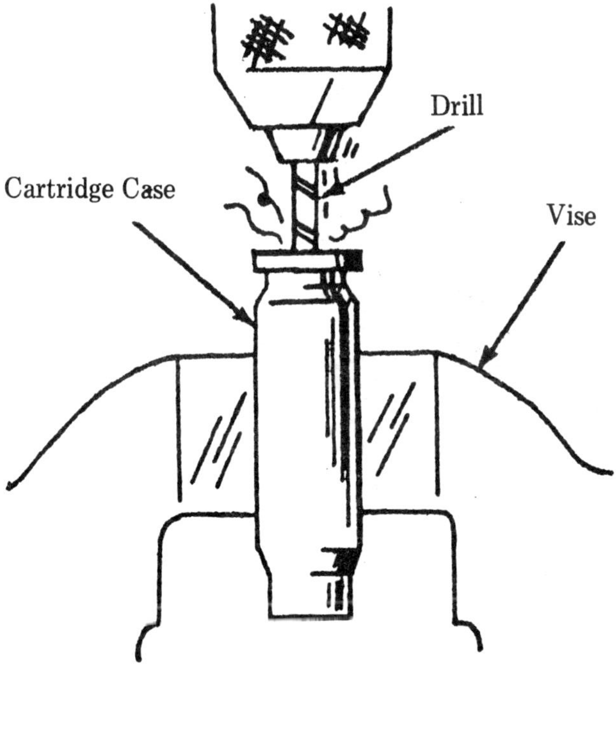

2) If necessary, open out flash hole in the primer pocket using a drill or knife. Make it large enough to receive a fuse.

3) Place one end of fuse in the flash hole and extend it through the case until it becomes exposed at the open end. Knot this end and then pull fuse in cartridge case thus preventing fuse from falling out.

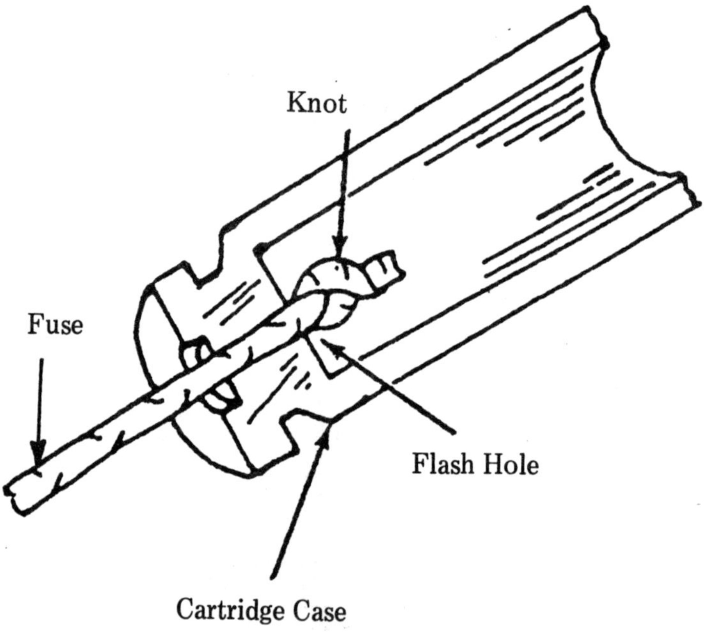

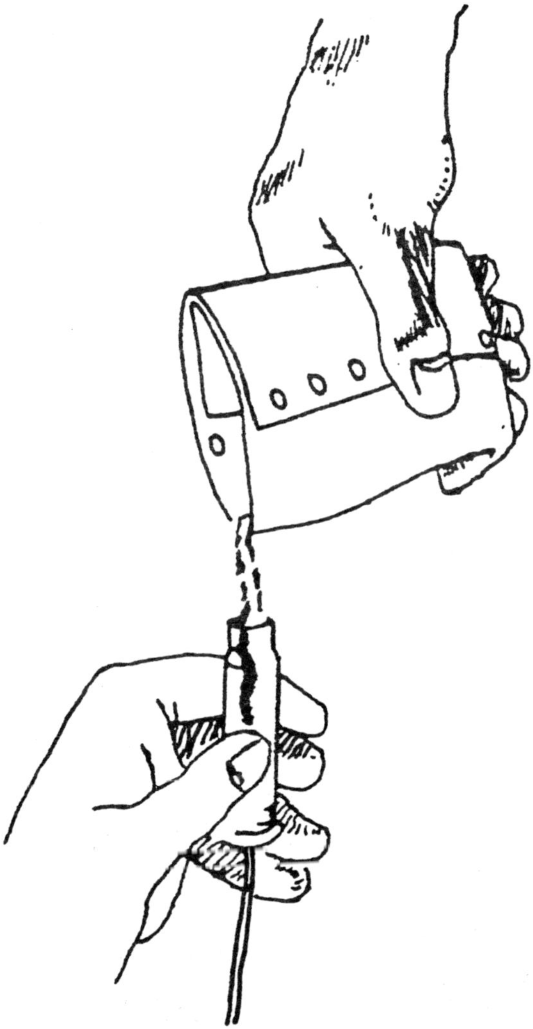

4) Load the Black Powder in the cartridge case.

5) Compress the Black Powder into the cartridge case with the wooden stick and the following improvised loading fixture.

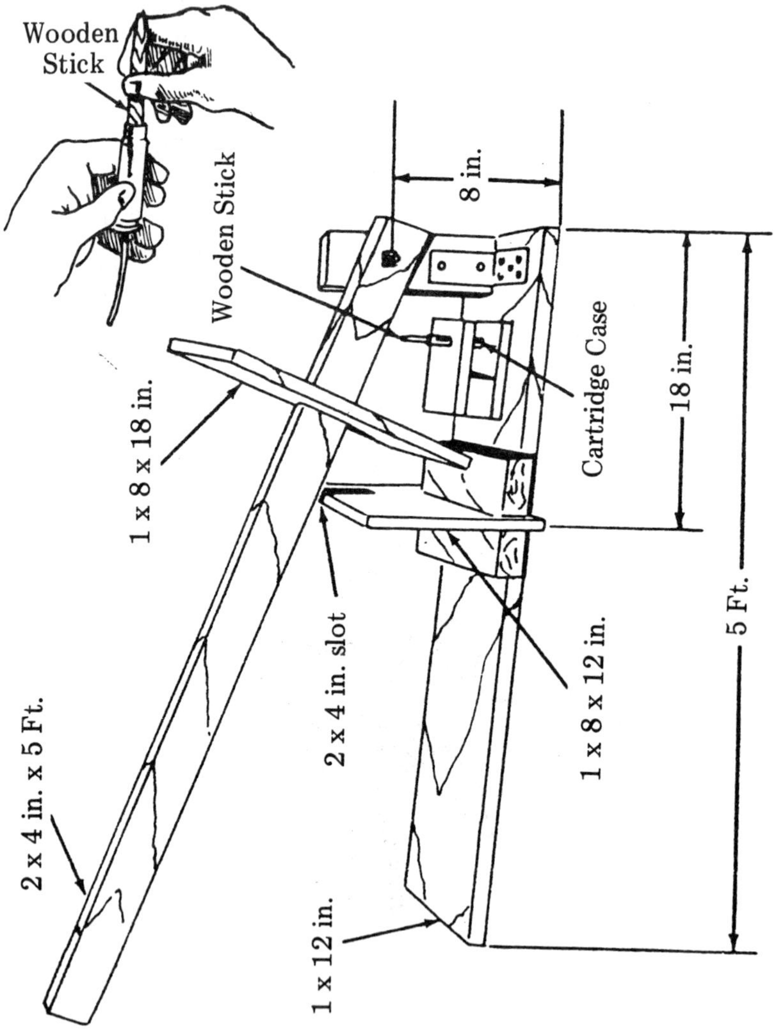

CAUTION: The explosive is shock and flame sensitive.

6) Fill the cartridge case with Black Powder until completely full.

7) If available, commercial reloading equipment could be used instead of improvised loading fixture.

CAUTION: These detonators have considerably more power than a military blasting cap and should be handled carefully.

Pipe Hand Grenade

Hand grenades can be made from a piece of iron pipe. The filler can be plastic or granular military explosive, improvised explosive, or propellant from shotgun or small arms ammunition.

MATERIAL REQUIRED:

Iron pipe, threaded ends, 1½ to 3 inches in diameter, 3 to 8 inches long

Two (2) iron pipe caps

Explosive or propellant

Non-electric blasting cap

Fuse cord

Hand drill

pliers

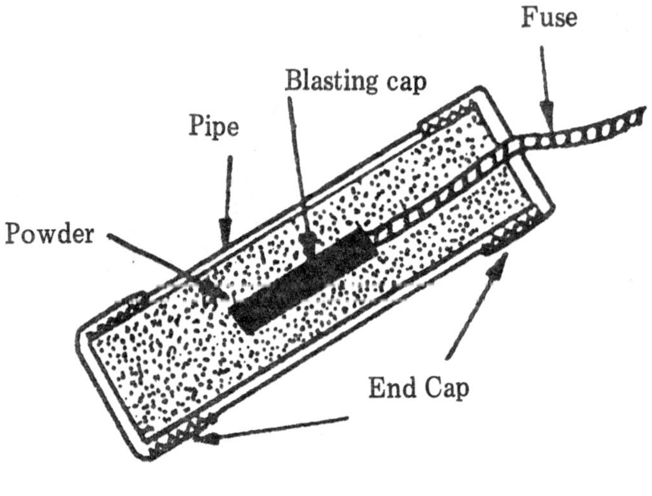

PROCEDURE:

1) Place blasting cap on one end of fuse cord and crimp with pliers.

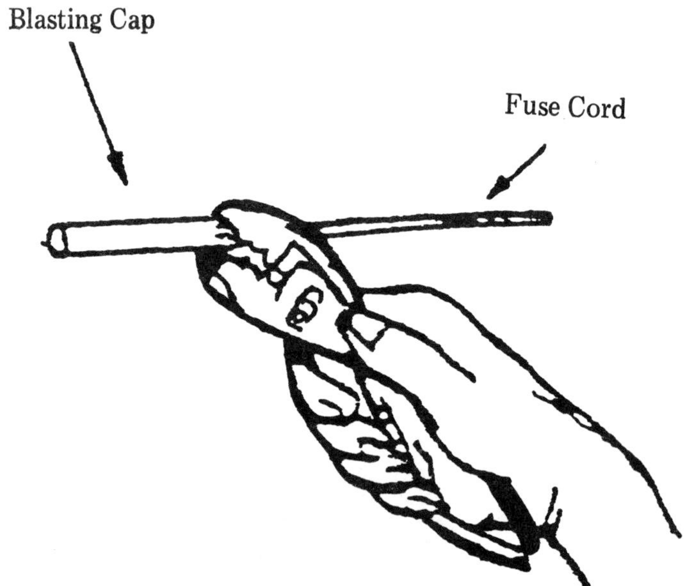

NOTE: To find out how long the fuse cord should be check the time it takes a known length to burn. If 12 inches burns in 30 seconds, a 6 inch cord will ignite the grenade in 15 seconds.

2) Screw pipe cap to one end of pipe. Place fuse cord with blasting cap into the opposite end so that the blasting cap is near the center of the pipe.

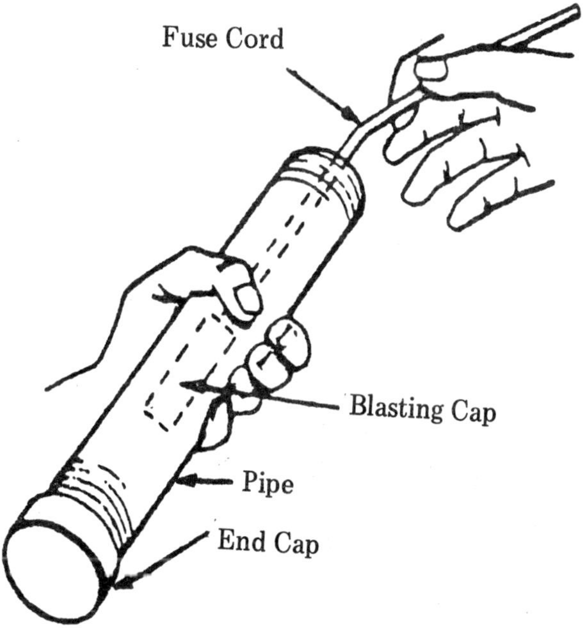

NOTE: If plastic explosvie is to be used, fill pipe BEFORE inserting blasting cap. Push a round stick into the center of the explosive to make a hole and then insert the blasting cap.

3) Pour explosive or propellant into pipe a little bit at a time. Tap the base of the pipe frequently to settle filler.

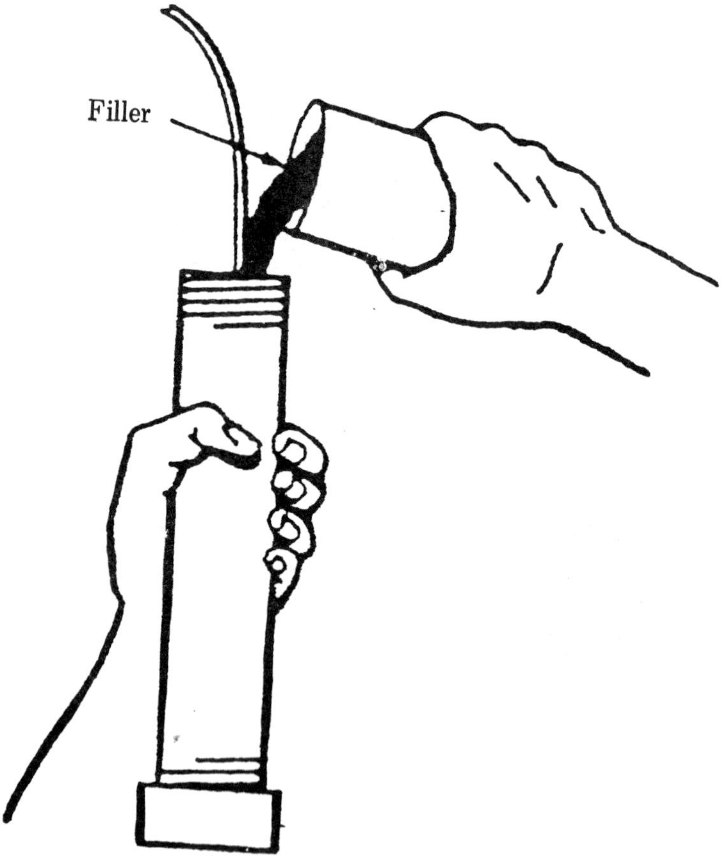

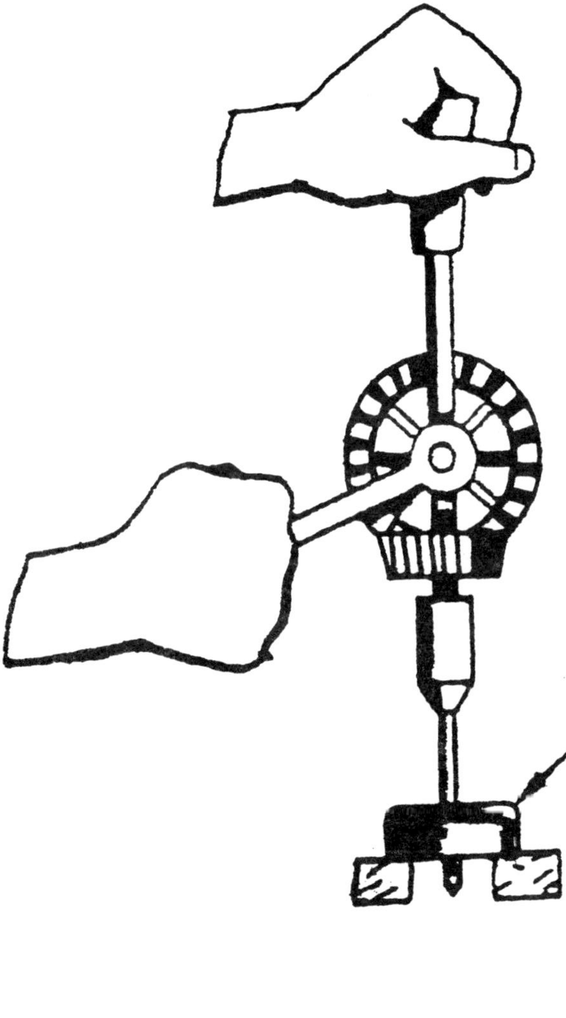

4) Drill a hole in the center of the unassembled pipe cap large enough for the fuse cord to pass through.

5) Wipe pipe threads to remove any filler material. Slide the drilled pipe cap over the fuse and screw handtight onto the pipe.

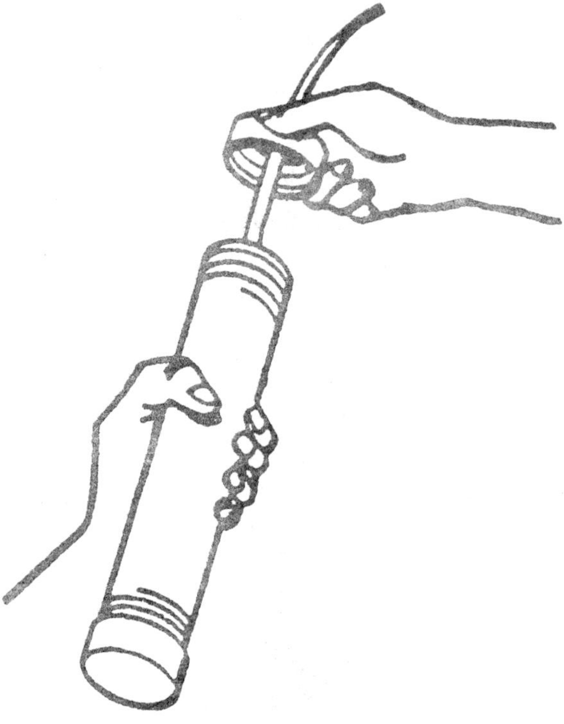

6) Pipe grenade is now ready to use.

Shotgun Grenade Launcher

This device can be used to launch a hand grenade to a distance of 160 yards (150 meters) or more, using a standard 12 gauge shotgun.

MATERIAL REQUIRED:

Grenade (improvised pipe hand grenade may be used)

12 gauge shotgun

12 gauge shotgun cartridges

Two washers, (brass, steel, iron, etc.), having an outside diameter of 5/8 inch

Rubber disk 3/4 inch in diameter and 1/4 inch thick (leather, neoprene, etc. can be used)

A 30 inch long piece of hard wood (maple, oak, etc.) approximately 5/8 inch in diameter. Be sure that the wood will slide into barrel easily.

Tin can (grenade and its safety lever must fit into can)

Two wooden blocks about 2 inch square and 1 1/2 inch thick

One wood screw about 2 inches long

12 gauge wads, tissue paper, or cotton

Adhesive tape, string, or wire

Drill

PROCEDURE:

1) Punch a hole in the center of the rubber disk large enough for a screw to pass through.

2) Make push-rod as shown.

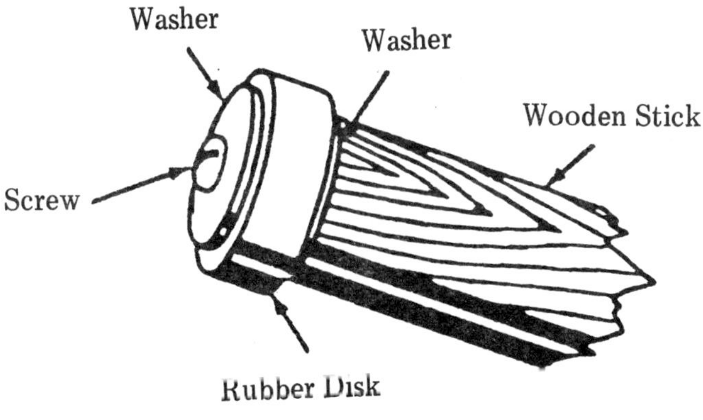

NOTE: Gun barrel is slightly less than ¾ inch in diameter. If rubber disk does not fit in barrel, file or trim it very slightly. It should fit tightly.

3) Drill a hole through the center of one wooden block of such size that the push-rod will fit tightly. Whittle a depression around the hole on one side approximately ⅛ inch and large enough for the grenade to rest in.

4) Place the base of the grenade in the depression in the wooden block. Securely fasten grenade to block by wrapping tape (or wire) around entire grenade and block.

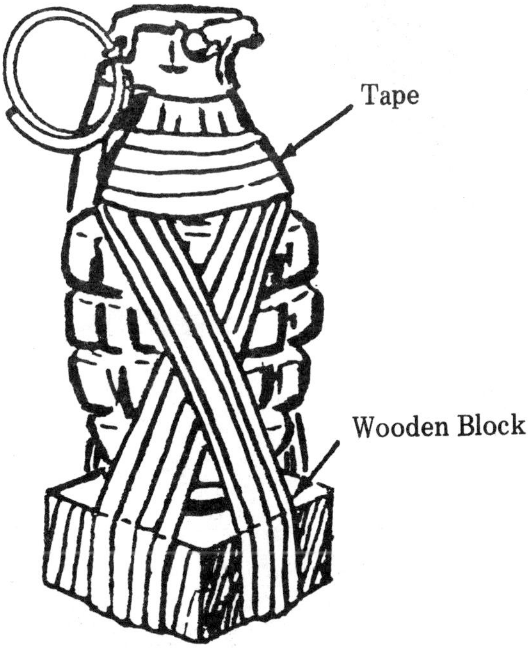

NOTE: Be sure that tape (or wire) does not cover hole in the block or interfere with the operation of the grenade safety lever.

5) Drill a hole through the center of the second wooden block, so that it will just slide over the outside of the gun barrel.

6) Drill a hole in the center of the bottom of the tin can the same size as the hole in the block.

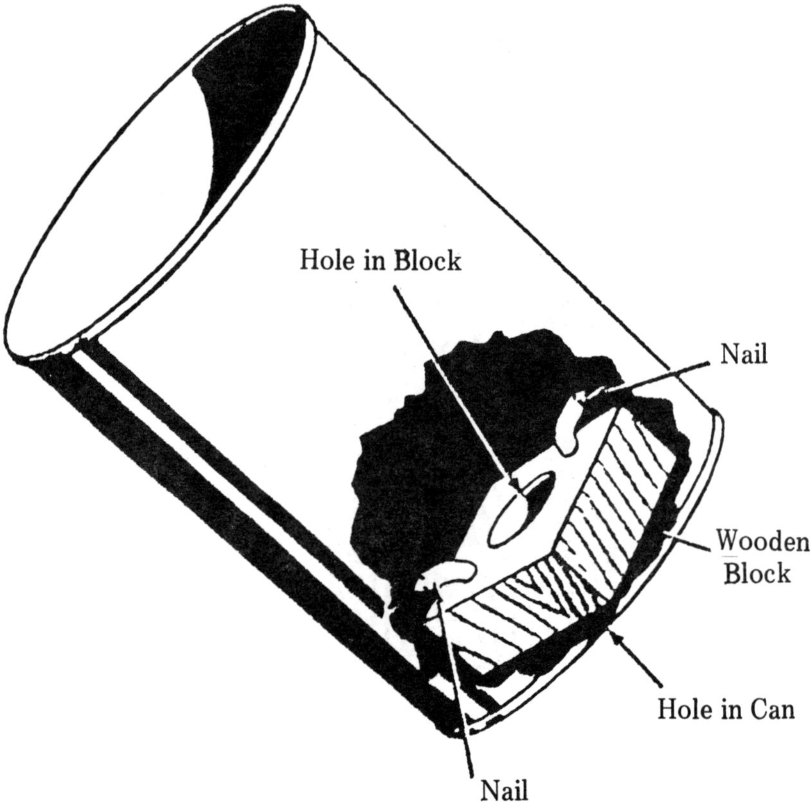

7) Attach can to block as shown.

8) Slide the can and block onto the barrel until muzzle passes the tin can's open end. Wrap a small piece of tape around the barrel an inch or two from the end. Tightly wrapped string may be used instead of tape. Force the can and wooden block forward against the tape so that they are securely held in place. Wrap tape around the barrel behind the can.

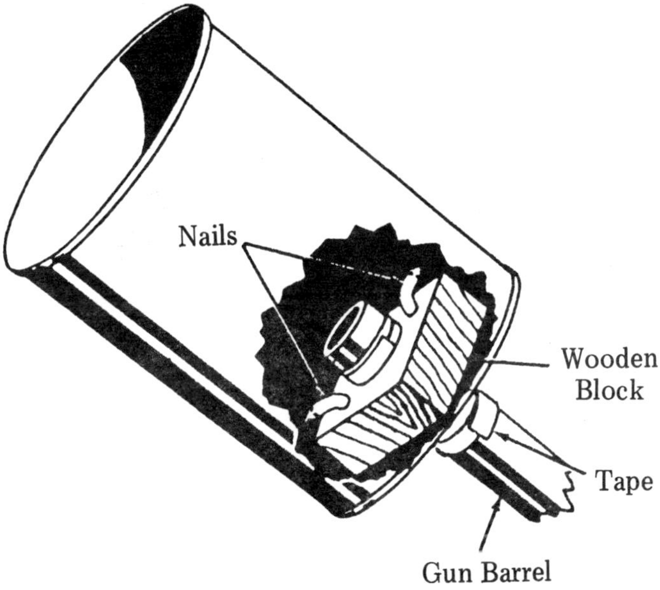

CAUTION: Be sure that the can is securely fastened to the gun barrel. If the can should become loose and slip down the barrel after launcher is assembled, the grenade will explode after the regular delay time.

9) Remove crimp from a 12 gauge shotgun cartridge with pen knife. Open cartridge. Pour shot from shell. Remove wads and plastic liner if present.

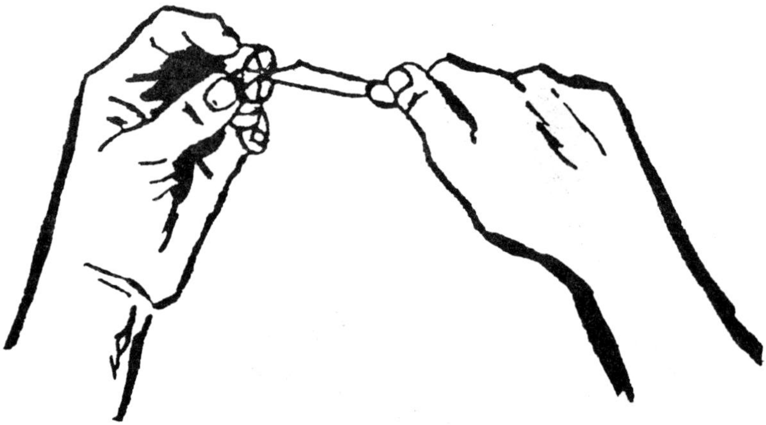

10) Empty the propellant onto a piece of paper. Using a knife, divide the propellant in half. Replace half of the propellant into the cartridge.

11) Replace the 12 gauge cardboard wads into cartridge case.

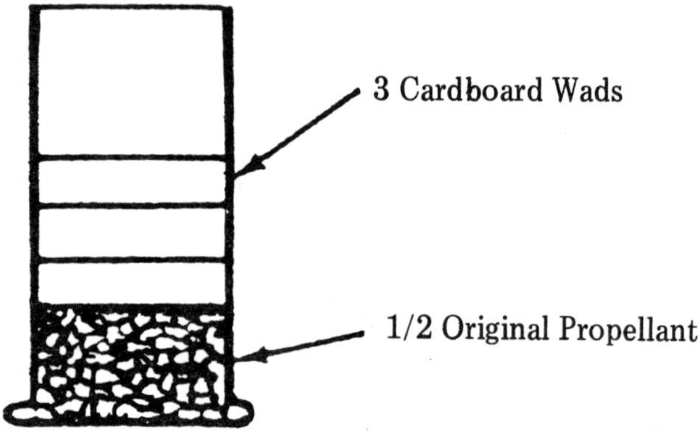

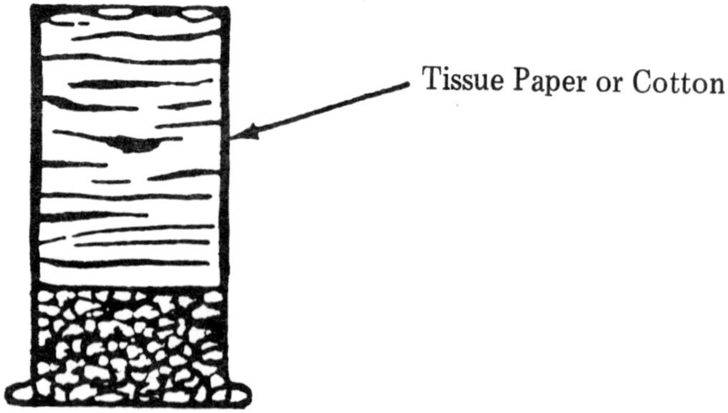

NOTE: If wads are not available, stuff tissue paper or cotton into the cartridge case. Pack tightly.

HOW TO USE:

METHOD 1 — When ordinary grenade is used:

1) Load cartridge in gun.

2) Push end of push-rod without the rubber disk into hole in wooden block fastened to grenade.

3) Slowly push rod into barrel until it rests against the cartridge case and grenade is in can. If the grenade is not in the can, remove rod and cut to proper size. Push rod back into barrel.

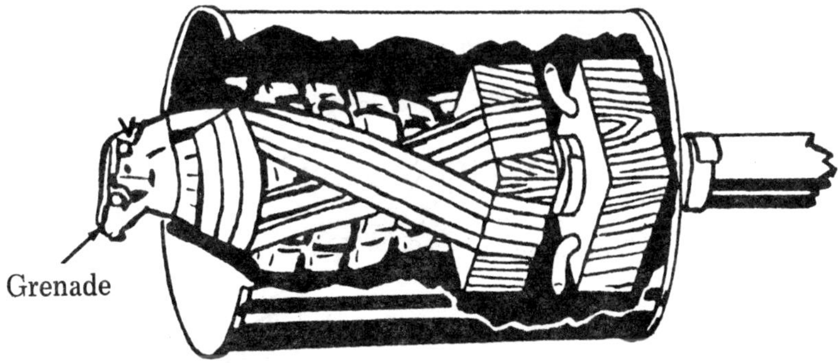

4) With can holding safety lever of grenade in place, carefully remove safety pin.

CAUTION: Be sure that the side of the can restrain the grenade safety lever. If the safety lever should be released for any reason, grenade will explode after regular grenade delay time.

5) To fire grenade launcher, rest gun on the ground at angle determined by range desired. A 45 degree angle should give about 160 yards.

METHOD 2 — When improvised pipe grenade is used:

An improvised pipe grenade may be launched in a similar manner. No tin can is needed.

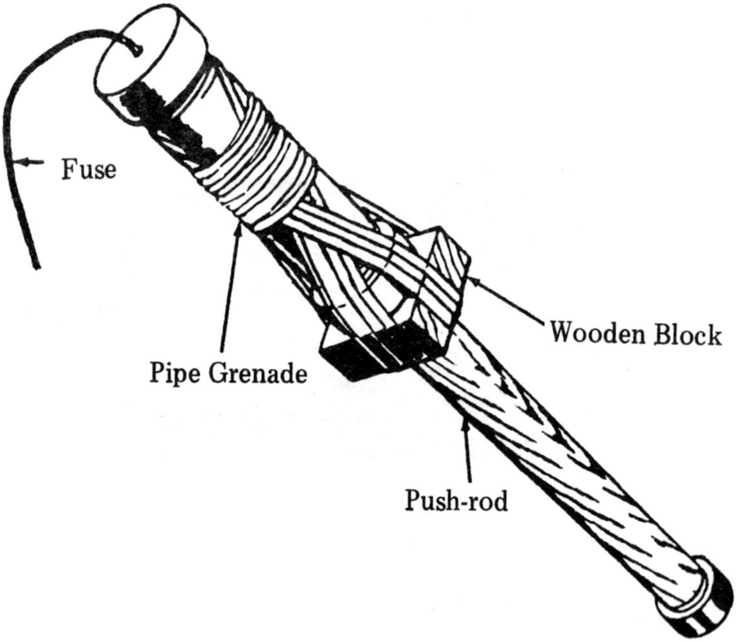

1) Fasten the grenade to the block as shown above with the fuse hole at the end opposite the block.

2) Push end of push-rod into hole in wooden block fastened to grenade.

3) Load cartridge in gun.

4) Push rod into barrel until it rests against the cartridge case.

5) Follow step 5 of method 1.

6) Using a fuse with AT LEAST a 10 second delay, light the fuse before firing.

7) Fire when the fuse burns to half its original length.

Improvised Claymore Mine

An improvised Claymore Mine can be easily constructed using common materials. Fused electrically, it is a command-detonating device designed for employment from ambush or defensive positions. It has a range of 150 to 200 meters, and is effective against personnel and thin-skinned vehicles.

MATERIAL REQUIRED:

Tin, metal, or cardboard sheet

Black powder or plastic explosive filler

Cloth or screen

Tape

Wooden sticks

Improvised electric blasting cap

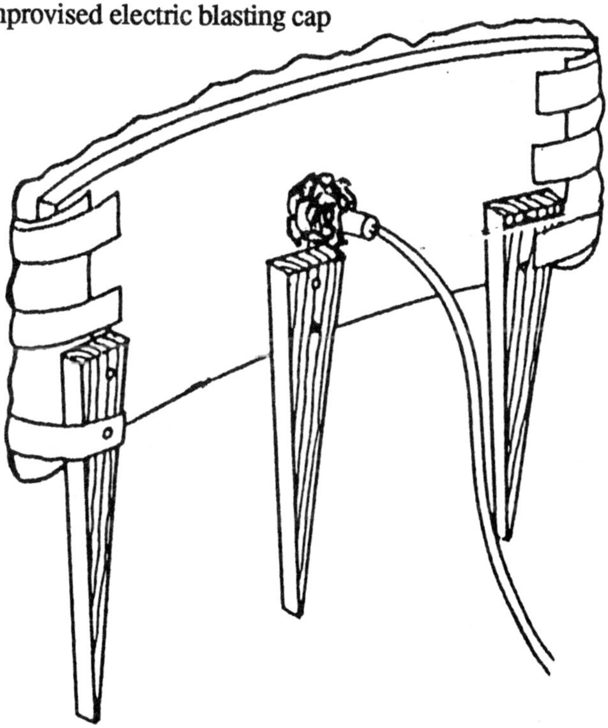

PROCEDURE:

1) Attach shrapnel to convex side of base and cover with cloth, tape, or screen retainer.

2) Place layer of explosive on concave side of base

3) Attach legs to concave side of base.

4) Attach electric blasting cap at exact rear center.

5) Attach firing device to firing wires at proper distance from mine for safety.

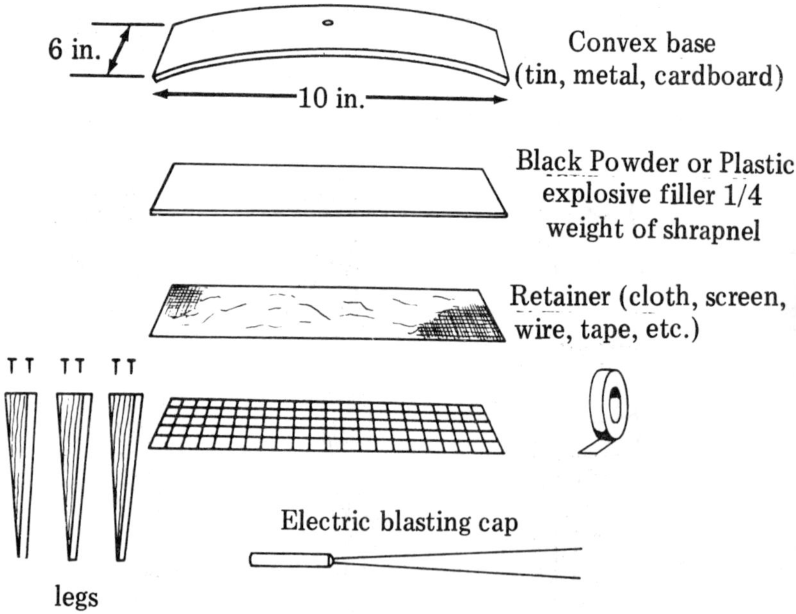

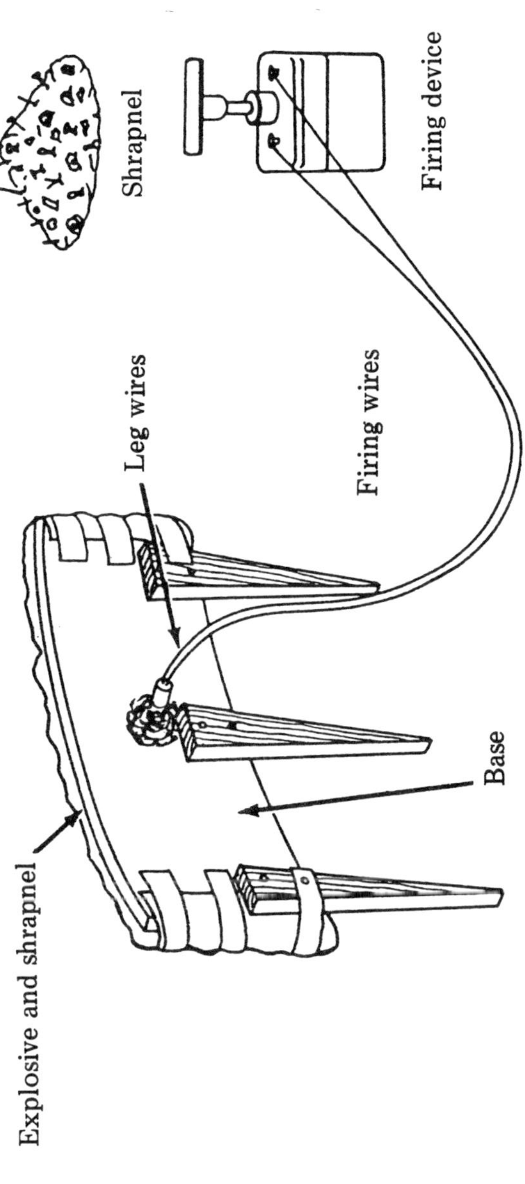
Shrapnel
Firing device
Leg wires
Firing wires
Base
Explosive and shrapnel

Plastic Explosive

Improvised Plastic explosive filler can be made from potassium chlorate and petroleum jelly (Vaseline). The potassium chlorate should be ground into very fine powder and then mixed with the petroleum jelly. This explosive can be detonated with commercial, military or improvised blasting caps. Store in a waterproof container until ready to use.

PROCEDURE:

Place 9 parts powdered potassium chlorate and 1 part petroleum jelly in a wide bowl or similar container. Mix ingredients with hands (knead) until a uniform paste is obtained.

Rocket Launcher

Making your own Rocket Launcher would seem to most people an impossible task, when in fact this device is one of the simpler of all improvised weapons to construct. There's no need for sophisticated machine tools—or difficult to obtain parts—as only common household tools and materials are used.

Rocket launchers are smooth bore, breech loading, single shot, shoulder held weapons of the open tube type. Firing electrically or mechanically ignited rockets, these rockets depend on their fins to keep them stable during flight. Rocket launchers can be fired from the shoulder in the standing, kneeling, sitting and prone positions.

In existence for over 40 years, earlier models were made of steel, and weighed about 15 pounds unloaded. Current military models are mostly made of plastic or fiberglass and are of the one-shot disposable type.

The design presented here is a very simple one to make, while still being highly effective. It should be noted that these devices are illegal to construct and to own, and are capable of causing devastating damage. Study, therefore, the information, but consult with the authorities before attempting to build any destructive device. All information is provided for academic study only!

The rockets presented here can be built from components available from model rocket companies. Parts are available at most hobby shops, toy stores or directly from the manufacturers.

These companies will send you catalogs that contain most of the parts necessary for the construction of bazooka rockets.

Estes Industries
Penrose, CO 81240

Centuri Engineering Co.
P.O. Box 1988
Phoenix, AZ 85001

LAUNCHER TUBE

MATERIALS NEEDED FOR CONSTRUCTION:

1) TUBE: Launcher tube should have a diameter of from 2 to 4 inches and approximately 4 to 6 feet in length. This tube can be made of steel, aluminum, plastic, fiberglass or cardboard. If using a cardboard tube, however, obtain one with a ⅛ to ¼ inch wall thickness. DO NOT use mailing tubes or gift wrapping paper tubes to build your launcher. Some plastic tubes are also not rigid enough, and won't remain straight when longer than 5 to 6 feet. To test rigidness, hold tube at the center and if it bends, DO NOT USE IT, as rocket performance will be erratic at best.

2) ELECTRONIC LAUNCHING DEVICE: A launching device is a switch that provides current to an igniter, thus launching the rocket. These devices are manufactured by model rocket companies. One of the better units on the market is called the Solar Launch Controller. Manufactured by Estes Industries of Colorado, this unit weighs just over 7 ounces, including the batteries and measures 6.1 inches in length. Height is only 2.4 by 1.44 inches in width.

Unit can be assembled in minutes and includes a safety key, continuity check lamp, ignition button (trigger device on your rocket launcher) and 15 feet of cable. Cost is around $15.

SAFETY CHECK

Follow the following procedures before attempting shoulder firing:

1) Locate a barrier such as a stone wall or large tree which you can stand behind in case the weapon explodes when fired.

2) Mount launcher solidly to a table or other rigid support at least ten feet in front of barrier.

3) Insert a rocket into the launcher tube and attach the alligator clips to the igniter in rocket engine.

4) Take control device and go behind the barrier. (The control device must be detached from tube.)

5) Push the firing buttom so that the weapon fires.

6) IMPORTANT: Fire at least five rounds from behind the barrier and then re-inspect the launcher before you attempt to shoulder fire it.

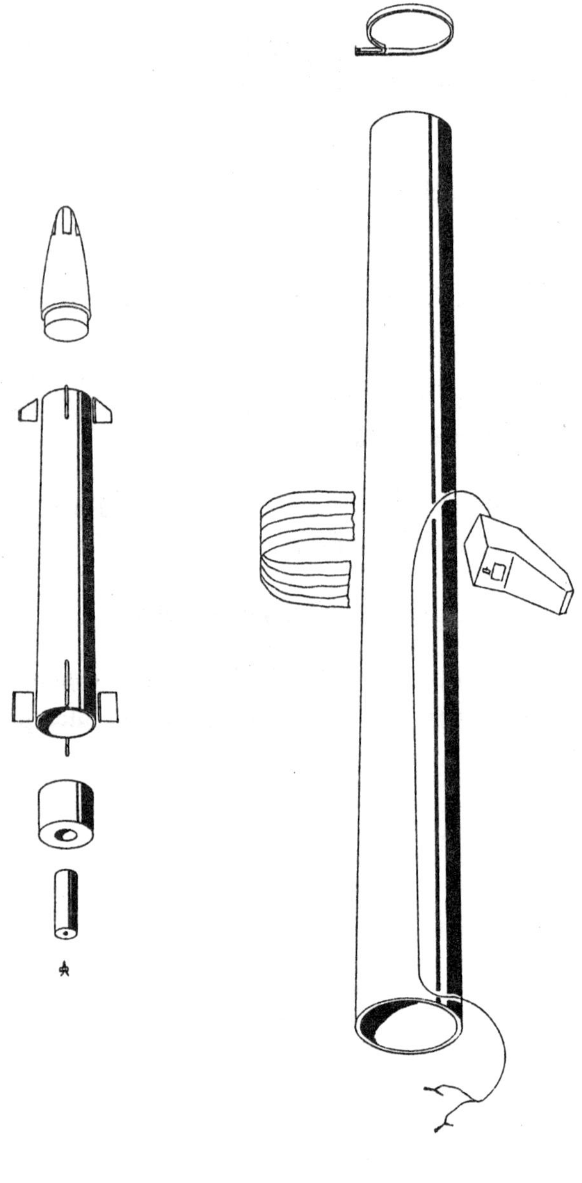

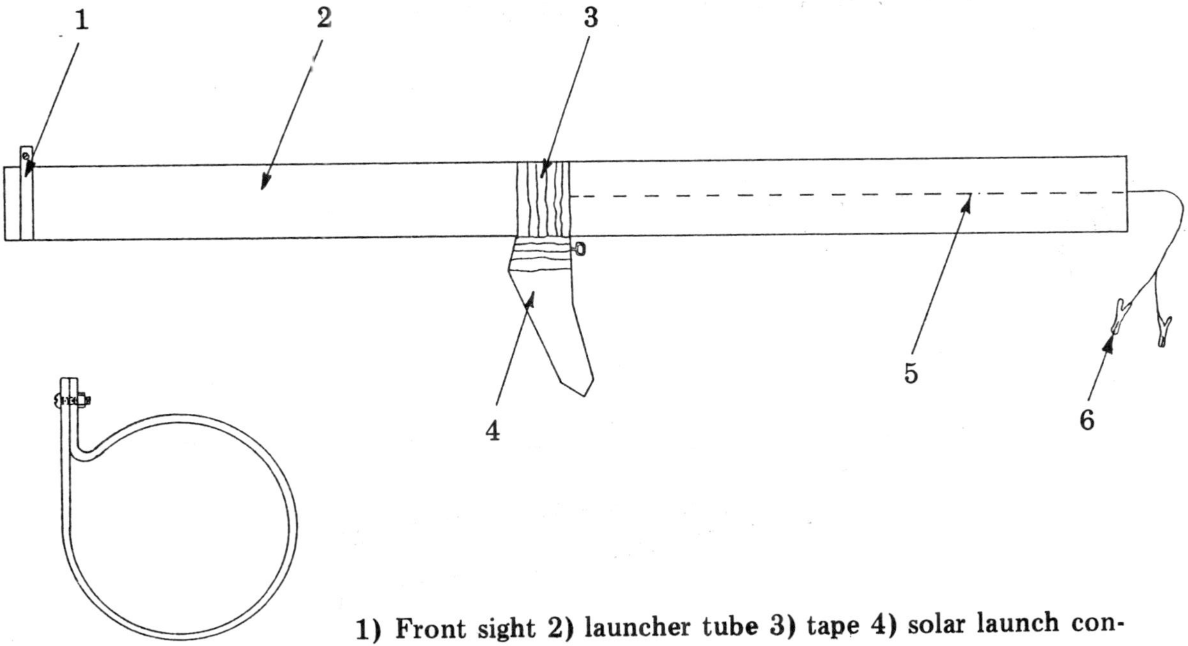

1) Front sight 2) launcher tube 3) tape 4) solar launch contoller 5) launch wire 6) alligator clips.

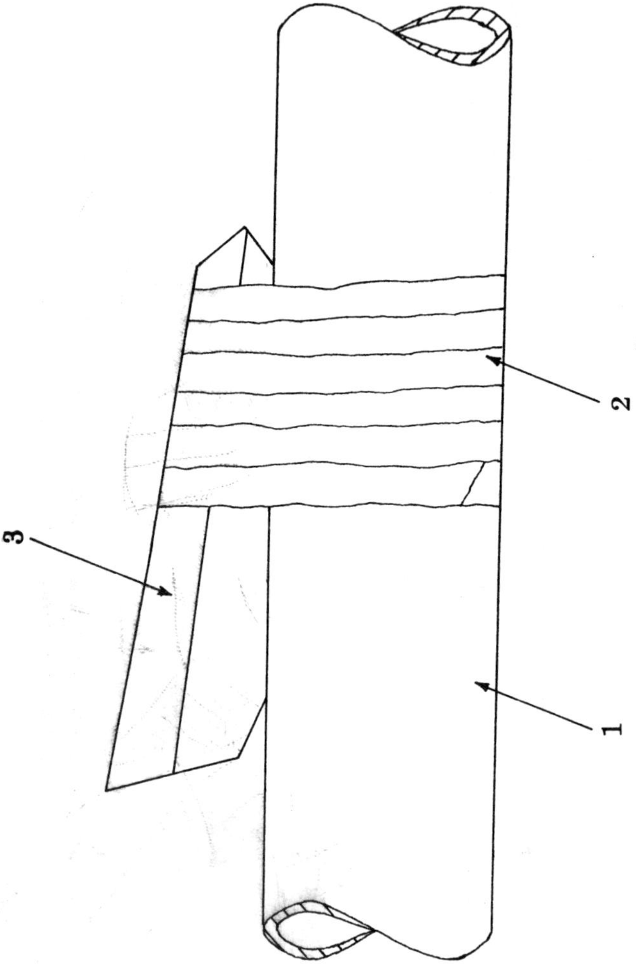

1) Launching tube 2) tape 3) solar launch control system.

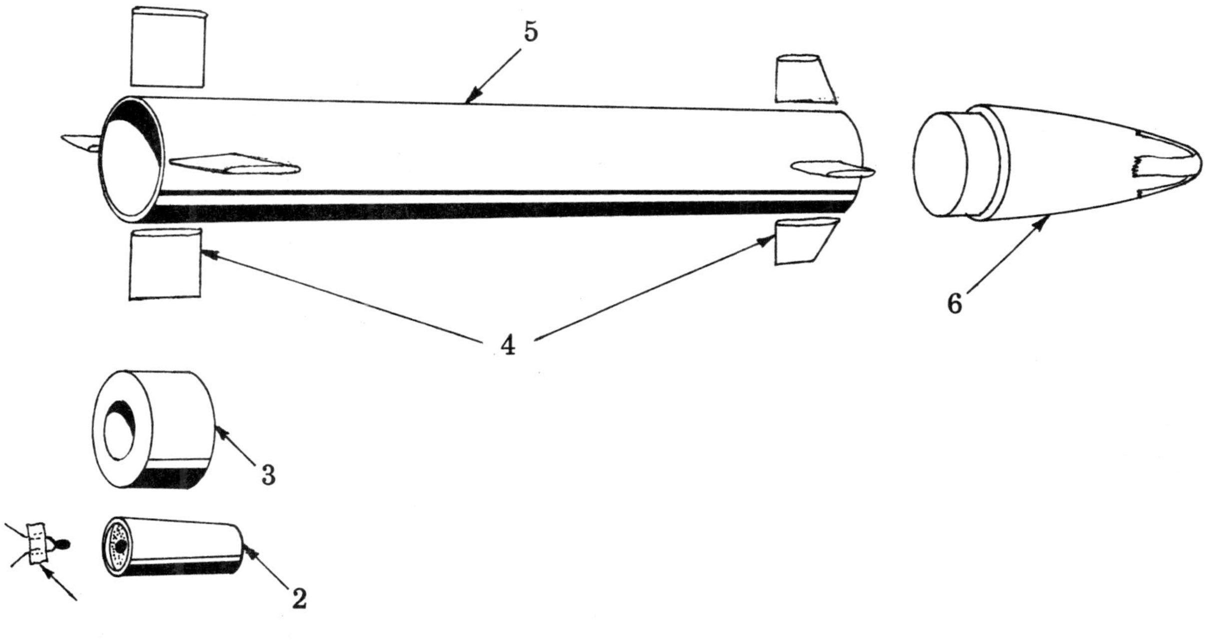

1) Model rocket engine igniter 2) rocket engine 3) engine block 4) balsa fins 5) rocket body tube (cardboard) 6) balsa nose cone.

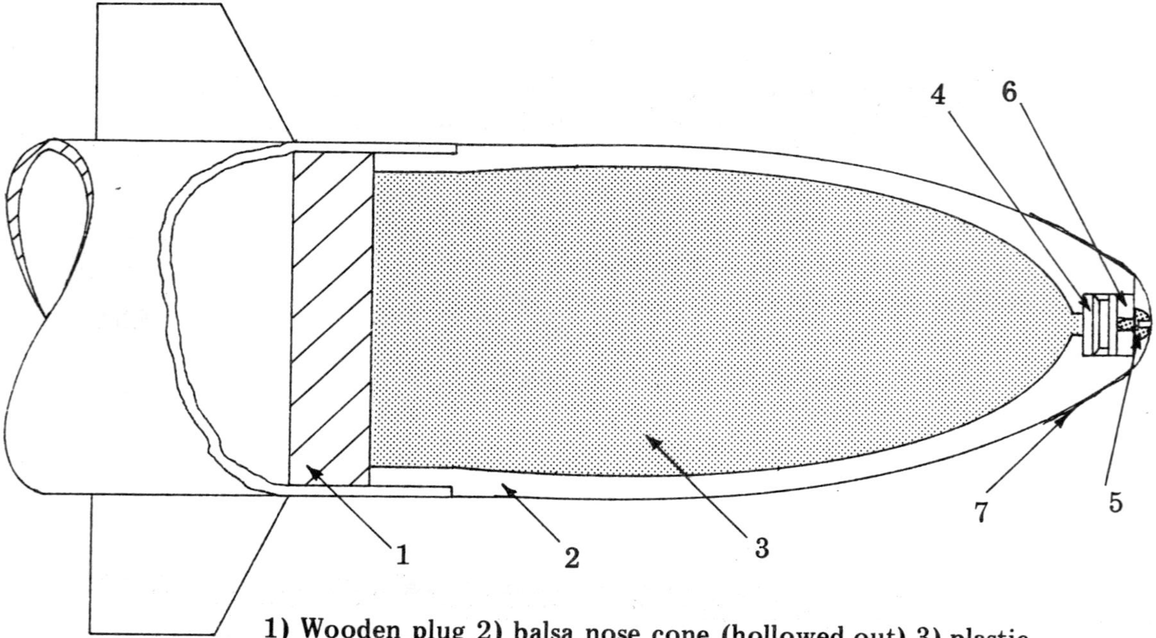

1) Wooden plug 2) balsa nose cone (hollowed out) 3) plastic filler explosive or black powder 4) primed pistol case 5) machine screw firing pin 6) styrofoam spacer (to hold firing pin in place) 7) electrical tape.